green black red

# green black red

## recipes for cooking and enjoying california grapes

BY SUSAN VOLLAND    FOREWORD BY SARA MOULTON    PHOTOGRAPHS BY E.J. ARMSTRONG

CHRONICLE BOOKS
SAN FRANCISCO

Library of Congress Cataloging-in-
Publication Data available.

ISBN 978-0-8118-6332-2


Manufactured in China

Cover and lead stylist: Diana Isaiou
Food styling by Patty Wittmann
Design by Alicia Nammacher
Recipes by Susan Volland
Text by James S. Howard
Typesetting by Connie Bigelow


Chronicle Books LLC
680 Second Street
San Francisco, California 94107

www.chroniclebooks.com

:::::contents

# foreword

Everyone loves grapes. That's easy. What is surprising to most people, however, is how versatile grapes are as an ingredient. As a professional chef and a home cook, I'm always looking for new ways to make meals interesting. When those ideas help turn flavorful food into healthy food, it's even better. In this book, you'll find dozens of inspiring ways that grapes work as an ingredient. They add color and sweetness to chicken salad and juicy flavor to grilled chicken skewers with rosemary, and are the ideal sweet touch at the end of a meal. And because grapes are so popular, you'll be making dishes that the whole family will enjoy. From cool grape poppers for the kids to an elegant grape cocktail for adults, this book is filled with new ways to enjoy the fruit of the vine. —Sara Moulton

# :::::: introduction

From easy to elegant, every meal is enhanced when you bring grapes to the table. Since they're always ripe and don't need peeling, coring, or seeding, fresh grapes are a simple addition to recipes. Grapes add color, texture, sweetness, and flavor without overpowering other ingredients. And since fresh grapes are one of the most popular foods in the world, you can add excitement to your cooking and still please even the pickiest eaters.

Grape and Prosciutto Appetizers

:::::::: start

:::::: Starters are a way to ease into a meal.  You get a few nibbles, but the conversation keeps going. For the cook, it's a chance to show off your style and have your guests try something new. Everyone knows that grapes are a fruit, and most people have heard that tomatoes are a fruit. But did you know that in almost any recipe where you can use a tomato, you can use a grape instead? When tomatoes aren't in season, add grapes to your green salad.  Instead of tomato-based gazpacho, try a classic green gazpacho made with fresh green grapes. These grape recipes start a meal with style.

Goat-Cheese-Stuffed Grapes
with Pistachios

# Green Grape Gazpacho

Nothing refreshes and revives quite like a chilled bowl of Spanish gazpacho. This vibrant variation has all the complexity of the classic tomato version, plus an appealing color and subtle sweetness.

MAKES
6 SERVINGS

| | |
|---|---|
| 2 2/3 | CUPS WASHED AND TORN ROMAINE LETTUCE |
| 1 | MEDIUM CUCUMBER, PEELED AND CHOPPED (1 3/4 CUPS) |
| 2 | LARGE CELERY STALKS, SLICED (1 CUP) |
| 1 | CUP GREEN GRAPES |
| 1 | CUP VEGETABLE BROTH |
| 3/4 | CUP SLICED GREEN ONIONS, BOTH GREEN AND WHITE PARTS |
| 1/3 | CUP CHOPPED GREEN BELL PEPPER |
| 1/3 | CUP CHOPPED DAY-OLD BREAD (CRUSTS REMOVED) |
| 1/4 | CUP SHERRY VINEGAR OR RED WINE VINEGAR |
| 2 | TABLESPOONS EXTRA-VIRGIN OLIVE OIL |
| 1 | TEASPOON CHOPPED GARLIC |
| 1/2 | TEASPOON SALT |
| 1/4 | TEASPOON WHITE PEPPER |
| | DICED TOMATO FOR GARNISH |

Combine all the ingredients, except the tomato, in a large bowl or container. Puree the soup in 2 or 3 batches in a blender or food processor until very smooth. Chill for at least 1 hour to improve the flavor.

Ladle the soup into bowls and garnish with a sprinkle of the diced tomato.

# Baby Greens with Grapes, Hot-Sweet Pecans, and Gorgonzola Vinaigrette

The universal appeal of this salad is the diversity of flavors and textures. It is crisp, crunchy, soft, sweet, spicy, and tangy.

| HOT-SWEET | ¼ | CUP BROWN SUGAR, LIGHTLY PACKED |
|---|---|---|
| PECANS | 2 | TABLESPOONS BUTTER |
| | ½ | TEASPOON CAYENNE PEPPER |
| SERVES 6 | ½ | TEASPOON SALT |
| | 1 | TEASPOON WATER |
| | 1 | CUP PECAN HALVES |
| | | |
| GORGONZOLA | 1 | CUP (4 OUNCES) CRUMBLED GORGONZOLA CHEESE |
| VINAIGRETTE | ½ | CUP EXTRA-VIRGIN OLIVE OIL |
| | 3 | TABLESPOONS WHITE WINE VINEGAR |
| | ½ | TEASPOON SALT |
| | ¼ | TEASPOON PEPPER |
| | | |
| | 6 | HANDFULS (6 OUNCES) MIXED BABY SALAD GREENS |
| | 2 | CUPS GREEN, RED, AND/OR BLACK GRAPES |

To make the pecans, first lightly grease a metal baking sheet and set aside. In a nonstick skillet or sauté pan, combine the brown sugar, butter, cayenne pepper, salt, and water. Bring to a boil over medium-high heat. Stir until the sugar is dissolved, about 2 minutes. Carefully add the pecan halves and stir constantly for one and a half minutes. The pecans should be well coated with the sugar mixture and crispy, but not scorched. Spoon the pecans onto the prepared baking sheet and spread to separate. Cool thoroughly.

To make the vinaigrette, combine the Gorgonzola cheese, olive oil, vinegar, salt, and pepper in a small bowl and whisk to blend.

To assemble the salad, toss the greens with the grapes, nuts, and vinaigrette. Serve immediately.

# Warm Spinach Salad with Grapes and Pancetta

A good spinach salad is always a welcome addition to your recipe collection, and this one is a winner. It makes an excellent starter to a seafood or pasta dinner, or a light main course all on its own.

SERVES 6

| | |
|---|---|
| 2 | BUNCHES FRESH SPINACH LEAVES, WASHED AND DRIED |
| 3 | TABLESPOONS BALSAMIC VINEGAR |
| ¼ | CUP EXTRA-VIRGIN OLIVE OIL |
| ¼ | TEASPOON SALT |
| ¼ | TEASPOON PEPPER |
| ½ | CUP (4 OUNCES) CHOPPED PANCETTA (OR SUBSTITUTE BACON IF PANCETTA IS UNAVAILABLE) |
| ¼ | CUP PINE NUTS |
| 1 | CUP GREEN, RED, AND/OR BLACK GRAPES |
| ½ | CUP THINLY SLICED RED ONION (ABOUT ¼ MEDIUM ONION) |
| ½ | CUP SHAVED OR GRATED PARMESAN CHEESE |

Place the spinach in a large bowl; set aside. In a small bowl, whisk together the balsamic vinegar, olive oil, salt, and pepper. Set aside. In a large sauté pan, cook the pancetta over medium-high heat until almost crisp, 5 to 7 minutes. Do not drain. Add the pine nuts to the pan with the pancetta and cook, stirring constantly, until they become lightly toasted, about 2 minutes. Add the grapes and red onion and continue to cook until warmed through. Pour the dressing over the pancetta mixture and bring to a boil. Pour the hot dressing over the prepared spinach. Toss quickly to coat the leaves. Arrange on a serving platter or individual plates. Sprinkle with the Parmesan shavings and serve.

# Grape and Brie Fritters

Yes, they are decadent, but these puffy fritters are like a divine revelation. Simply stir chunks of ripe Brie and fresh sweet grapes into choux pastry and deep-fry. Serve warm in a basket alongside a bottle of wine for a pre-event nosh or as a prelude to a bistro-style dinner.

MAKES ABOUT
40 TWO-INCH
FRITTERS

| | |
|---|---|
| 1 | CUP WATER |
| ½ | CUP (1 STICK) BUTTER, CUT INTO ½-INCH DICE |
| 1 | CUP ALL-PURPOSE FLOUR |
| 8 | OUNCES BRIE CHEESE, DICED |
| 3 | EGGS, BEATEN |
| ¼ | TEASPOON SALT, PLUS MORE FOR SPRINKLING |
| 2 | CUPS RED GRAPES, HALVED |
| | VEGETABLE OR PEANUT OIL FOR FRYING |

Put the water and diced butter in a medium saucepan and cook gently until the butter has melted. Increase the heat and bring to a rolling boil. Remove the pan from the heat and pour in all of the flour at once. Stir with a wooden spoon until the mixture is smooth and comes away from the sides of the pan. Let cool slightly.

Place the warm mixture in the bowl of a mixer. Add the diced Brie and mix until well combined. In a small bowl, whisk together the egg and salt and gradually add to the dough, beating well between each addition. The dough will be glossy and smooth. Stir in the grapes.

Carefully drop 5 to 6 spoonfuls of batter into 350-degree-F oil. Cook, turning occasionally, for 3 minutes, or until the fritters are completely puffed and a dark golden brown. Drain well on crumpled paper towels and repeat with remaining batter. Serve warm with a sprinkling of salt.

# Grilled Antipasto Skewers

These skewers are a tidy and innovative way to serve the classic ingredients of a good antipasto. When weaving the prosciutto throughout the skewer, try to resist the urge to press the ingredients down to the bottom of the skewer after you add them. Simply add an ingredient, weave the prosciutto over it, add another ingredient, and repeat. If the process seems too fussy for you, chunks of salami make a good alternative.

| | | |
|---|---|---|
| MARINADE | 1 | CLOVE GARLIC, CRUSHED |
| | 1 | TABLESPOON RED WINE VINEGAR |
| MAKES | 1/2 | TEASPOON CHOPPED FRESH ROSEMARY (1/4 TEASPOON DRIED) |
| 12 SKEWERS | 1/4 | TEASPOON SALT |
| | | FRESHLY GROUND BLACK PEPPER |
| | 1/4 | CUP EXTRA-VIRGIN OLIVE OIL |
| | | |
| | 12 | LONG BAMBOO SKEWERS |
| | 24 | PEELED AND DEVEINED SHRIMP |
| | 2 | CUPS GREEN, RED, AND/OR BLACK GRAPES |
| | 1 | (8 OUNCE) JAR COCKTAIL ONIONS |
| | 4 | SLICES (3 OUNCES) PROSCIUTTO, EACH PIECE SLICED INTO 3 LONG RIBBONS |

To make the marinade, mix the garlic, vinegar, rosemary, salt, and pepper in a small bowl and gradually whisk in the oil.

Thread the skewers with the shrimp, grapes, and onions, with a ribbon of prosciutto winding throughout. Place the finished skewers in a shallow dish or plate. Evenly dab on half of the marinade with a brush. Let marinate in the refrigerator for at least 30 minutes or as long as 24 hours. Brush with the remaining marinade just before cooking.

Grill the antipasto skewers over a hot fire for 2 minutes a side. (Alternatively, broil the skewers under a preheated broiler for 2 minutes per side.) Serve hot.

# Focaccia with Grapes and Rosemary

Chewy, fresh focaccia is as easy to make as homemade pizza. Traditionally, it was made with wine grapes, but table grapes are always available and you don't have to worry about the seeds. This version is baked in a depth of olive oil for traditional taste and texture. If you prefer, just lightly oil the baking pan to prevent sticking.

SERVES 8

| | |
|---|---|
| 1 | ENVELOPE (¼ OUNCE OR ¾ TABLESPOON) DRY YEAST |
| 2 | CUPS WARM (110 DEGREES F) WATER |
| 1 | TABLESPOON SUGAR |
| 5 TO 6 | CUPS UNBLEACHED ALL-PURPOSE FLOUR |
| 1 | TEASPOON SALT |
| ½ | CUP EXTRA-VIRGIN OLIVE OIL |
| 1½ | CUPS GREEN, RED, AND/OR BLACK GRAPES |
| 2 | TEASPOONS CHOPPED FRESH ROSEMARY (1 TEASPOON DRIED) |
| | COARSE SALT |

In a small bowl, sprinkle the yeast onto the water and stir together with the sugar. Let the mixture rest until the yeast blooms, 3 to 4 minutes.

If using an electric stand mixer, combine the yeast mixture, 3 cups of the flour, and the salt. Mix well, then change to the dough hook and gradually add more of the flour, kneading well between each addition, until the dough is smooth, fairly firm, and no dough sticks to the side of the bowl, about 5 minutes. If not using a mixer, stir together the yeast mixture, 3 cups of flour, and the salt and mix together with a large wooden spoon. Turn the dough out onto a heavily floured board and knead while gradually incorporating more flour into the dough until it is smooth and elastic, about 10 minutes.

Place the dough in a lightly oiled bowl and cover with plastic wrap or a dry cloth. Let rest in a warm place until the dough has doubled, about 1 hour.

Preheat oven to 425 degrees F.   >>>

Focaccia with Grapes and Rosemary
( CONTINUED )

Pour all but 2 tablespoons of the olive oil onto a 12-by-16½-inch baking sheet or 2 pizza pans. Simply lift the dough from the bowl and gently stretch and press it to fit the pan. Drizzle the dough with the remaining olive oil and dimple the top of the bread with your fingertips. Press the grapes into the dough evenly all over the bread, leaving about 1 inch between grapes. Sprinkle the bread generously with the chopped rosemary and coarse salt.

Bake the focaccia until it is a nice crisp brown, about 30 minutes. Remove from the oven and cut with a pizza cutter into squares. Serve warm with cheese and wine, or use to make into wonderful panini sandwiches.

## Goat-Cheese-Stuffed Grapes with Pistachios
( PICTURED ON PAGE 15 )

Preparing these is delicate work, but the dainty, flavor-filled tidbits are well worth the effort. Serve them as a spectacular passed appetizer or perhaps as a charming *amuse-bouche* before a multicourse dinner.

| MAKES ABOUT | 3 | CUPS (1 POUND) GREEN, RED, AND/OR BLACK GRAPES |
| 30 STUFFED | 8 | OUNCES SOFT GOAT CHEESE |
| GRAPES | 1 | CUP PISTACHIOS, CHOPPED |

Cut each grape in half and scoop out part of the flesh with a small melon baller. Roll the goat cheese into small balls and place in the cavity of the grapes. Sprinkle with the chopped pistachios.

Chill until ready to serve.

# Grape and Prosciutto Appetizers
( PICTURED ON PAGE 12 )

These crostini make an elegant party appetizer or alfresco snack before a Mediterranean-themed dinner.

| MAKES | | |
|---|---|---|
| 12 APPETIZERS | 1 | BAGUETTE, CUT INTO ¼-INCH SLICES |
| | 4 | TABLESPOONS OLIVE OIL |
| | 36 | GREEN, RED, AND/OR BLACK GRAPES |
| | 1 | TABLESPOON BALSAMIC VINEGAR |
| | | FRESHLY GROUND BLACK PEPPER |
| | 1 | BUNCH FRESH BASIL |
| | 4 | OUNCES SOFT GOAT CHEESE |
| | 6 | SLICES PROSCIUTTO |

Preheat oven to 350 degrees F. Lightly brush 12 slices of baguette with half of the olive oil and lay them directly on the middle rack of the oven. Toast for 5 minutes, or until light golden brown. Put the remaining olive oil in a small sauté pan and warm over medium-high. Add the grapes and sauté for 2 to 3 minutes, or until soft and warmed through. Remove from the heat. Sprinkle the balsamic vinegar over the grapes and add a few grindings of pepper.

Top each baguette slice with one leaf of basil and spread the goat cheese evenly among the toasted slices. Cut each slice of prosciutto in half and drape over the cheese. Top each of the baguettes with three of the sautéed grapes.

May be served warm or made in advance and served at room temperature.

Great Grape Smoothie

::::::: drink

Wine is certainly the most popular way
to drink grapes, but it's not the only way.
Freeze grapes and use them as ice cubes
to chill your drink without watering it down,
or make some fresh grape juice for a refreshing
drink everyone in the family will enjoy.

California Grape Cocktail

Sangria

# Sangria

A classic fruit and wine punch from Spain. This can be served in goblets, like wine, or over ice in tumblers with the added sparkle of soda water.

SERVES 4

| | |
|---|---|
| 1 | BOTTLE (750 ML) FULL-BODIED, DRY RED WINE |
| 2 | CUPS RED OR BLACK GRAPES |
| 1 | ORANGE, CUT INTO ¼-INCH SLICES |
| 1 | NECTARINE OR PEACH, CUT INTO THIN WEDGES |
| 2 | TABLESPOONS HONEY |
| 1 | CUP SODA WATER (OPTIONAL) |
| 8 | SMALL CLUSTERS OF GREEN GRAPES |

Combine all the ingredients except the soda water and the grape clusters in a pitcher. Stir well. Refrigerate for 2 to 24 hours to allow the flavors to meld. Add soda water if desired and serve with or without the fruit pieces, in glasses garnished with small clusters of grapes.

# Grape Daiquiris

Sometimes, nothing quite hits the spot like a blended fruit drink. This slushy combination of fresh grapes, lime, and white rum makes the perfect cooler for a hot summer day. To make a nonalcoholic version, simply omit the rum.

MAKES 2 DRINKS

| | |
|---|---|
| 1 | CUP GREEN, RED, OR BLACK GRAPES |
| 2 | JIGGERS (4 OUNCES) WHITE RUM |
| 2 | SHORT JIGGERS (3 OUNCES) SWEETENED LIME JUICE (SUCH AS ROSE'S) |
| 3 | CUPS ICE |

Place all the ingredients into the bowl of a blender and puree until blended and slushy, about 2 minutes. Serve in tall glasses.

# Great Grape Smoothie

( PICTURED ON PAGE 28 )

Don't forget that a handful of grapes adds a great boost of fiber, antioxidants, and vitamins to any fruit smoothie.

| MAKES | | |
|---|---|---|
| 2 DRINKS | 1 | CUP SEEDLESS GRAPES |
| | ½ | CUP FROZEN CHERRIES |
| | ½ | CUP FROZEN STRAWBERRIES |
| | ½ | CUP PEELED AND SLICED ORANGE |
| | ½ | CUP SLICED BANANA |

Combine all the ingredients in the jar of a blender. Blend until smooth.

Pour into two (12-ounce) glasses.

## FROZEN GRAPES AND GRAPE RINGS

Frozen grapes are a staple in many health-conscious households. They are a good alternative when you are craving sweets or ice cream, and they make the perfect snack. A handful of these "polar berries" dropped in a beverage, even a glass of water, will bring instant smiles. O To dress up a bowl of your favorite punch, make decorative ice garnishes by filling a ring mold or Bundt pan with an assortment of grapes and just enough water to cover. Freeze until hard and turn out into a punch bowl, letting the loose, frozen grapes scatter into the punch as it dissolves.
O Can be made in smaller molds and ice cube trays as well.

# Bubble Milk Tea

Funky shops serving "bubble tea" started in Taiwan and can now be found in cities all over the world. The tall, cold tea drinks are traditionally made with a strong brewed tea mixed with brown sugar syrup and big, black, chewy tapioca balls that fall to the bottom of the drink. Since the tapioca balls can be tricky to find and don't appeal to everyone, we suggest using fresh grapes as an alternative! Serve the drinks with long spoons, or look for the superfat, colorful Bubble tea drink straws at Asian markets or restaurant-supply stores.

| MAKES ONE | ½ | CUP BLACK OR RED GRAPES |
|---|---|---|
| 20-OUNCE | ½ | CUP MILK |
| GLASS OF | ¼ | CUP VERY STRONG BREWED BLACK, GREEN, OR JASMINE TEA (RECIPE FOLLOWS) |
| BUBBLE TEA | 2 | TABLESPOONS BROWN SUGAR SYRUP (RECIPE FOLLOWS, OR SUBSTITUTE A FLAVORED ITALIAN COFFEE/SODA SYRUP OF YOUR CHOICE) |
| | | ICE TO FILL THE GLASS |

| TEA | 1 | QUART BOILING WATER |
|---|---|---|
| | 6 TO 8 | TEABAGS OF BLACK, GREEN, OR JASMINE TEA, OR 6 TO 8 TABLESPOONS LOOSE TEA |

To brew a batch of very strong tea, add the teabags or tea leaves to the boiling water. Strain and cool. Store at room temperature.

| BROWN SUGAR | ½ | CUP BROWN SUGAR, LIGHTLY PACKED |
|---|---|---|
| SYRUP | ½ | CUP WHITE SUGAR |
| | 1 | CUP WATER |

Stir together the sugars and water in a small saucepan. Boil for five minutes and then cool. Store in the refrigerator for up to 2 weeks.

To prepare the drink, pour the grapes, milk, strong tea, and brown sugar syrup into a 20-ounce glass and stir well. Fill with ice and serve immediately.

# Grape and Green Tea Smoothie

This drink is rich in antioxidants. If green tea powder is unavailable, substitute ½ cup tepid brewed green tea plus 1 teaspoon sugar for the water and green tea powder.

MAKES 1 DRINK

| | |
|---|---|
| 1 | CUP GREEN GRAPES |
| ½ | CUP COLD WATER |
| ½ | CUP DICED PINEAPPLE (USE FRESH, CANNED, OR FROZEN) |
| ½ | CUP ICE CUBES |
| 1 | TEASPOON SWEETENED GREEN TEA POWDER |

Puree all ingredients in a blender until smooth. Pour into a tall glass and serve immediately.

# California Grape Cocktail
( PICTURED ON PAGE 31 )

This bright and refreshing cocktail is made with fresh grapes, lemon juice, and vodka with just a splash of champagne. It's perfect before dinner or with a crowd for movie or book club nights.

MAKES 2 DRINKS

| | |
|---|---|
| 2 | CUPS GREEN, RED, OR BLACK GRAPES |
| | JUICE OF 1 LEMON (3 TABLESPOONS) |
| 2 TO 4 | OUNCES VODKA |
| ¼ | CUP SUGAR |
| 2 | OUNCES CHAMPAGNE |

Blend all but one grape with the lemon juice. Strain through a fine-mesh sieve. Measure 2 parts juice to 1 part vodka into a small pitcher or measuring cup. Spread the sugar onto a small plate. Slice the reserved grape in half and use it to moisten the rims of two martini glasses. Dip the rims in the sugar and shake off any excess. Divide the juice and vodka mixture evenly between the glasses and top off with a splash of champagne. Serve immediately.

Grape and
Green Tea
Smoothie

Pantry Stuffed French Toast

:::::: entertain

For more than 8,000 years, people have been entertaining with grapes. That's because you can't eat grapes without sharing. Sharing a peach is messy. Sharing a banana is odd. But sharing grapes is natural. So when grapes were first cultivated in 6000 B.C., early humans invariably shared. Today, you can share grapes with your guests by putting them in a bowl on the table or, better yet, making Seared Duck Breast with Black Grapes and Port Reduction. The following recipes are stunning examples of ways that fresh grapes add flavor, color, and contrast to elegant meals.

Mediterranean
Grilled Chicken
and Grape Skewers

# California Wine Country Salads

In the spirit of the classic French salade niçoise, we have gathered the best flavors of the California wine country and assembled them in one composed-style entrée salad.

SERVES 6

| | | |
|---|---|---|
| 12 | HANDFULS (12 OUNCES) TENDER SALAD GREENS |
| 1½ | CUPS (6 OUNCES) TOASTED WHOLE ALMONDS |
| 3 | CUPS (1 POUND) GREEN, RED, AND/OR BLACK GRAPES |
| 6 | THICK, BACON SLICES, COOKED CRISP AND CHOPPED |
| 3 | CHICKEN BREASTS, GRILLED, CHILLED, AND DICED |
| 2 | RIPE AVOCADOS, CUBED |
| 6 | OUNCES GOAT CHEESE, CRUMBLED |
| | ZINFANDEL VINAIGRETTE (RECIPE FOLLOWS) |

Arrange the greens in 6 large plates. Arrange each of the other ingredients in separate clusters on the greens. Serve with small pitchers of the vinaigrette alongside.

ZINFANDEL
VINAIGRETTE
(ZIN VIN)

MAKES 2 CUPS

| | | |
|---|---|---|
| 1 | CUP CALIFORNIA ZINFANDEL |
| 1 | TABLESPOON SUGAR |
| ¼ | CUP LEMON JUICE |
| 1 | MEDIUM SHALLOT, CHOPPED |
| 1 | LARGE CLOVE GARLIC, CHOPPED |
| 1 | TEASPOON DRY MUSTARD |
| 1 | TEASPOON SALT |
| ½ | TEASPOON PEPPER |
| ½ | TEASPOON DRIED THYME (1½ TEASPOONS FRESH) |
| ½ | CUP EXTRA-VIRGIN OLIVE OIL |
| ½ | CUP VEGETABLE OIL |

In a nonreactive saucepan, boil the red wine until it has reduced to ½ cup. Cool.
In a blender, puree the reduced wine, sugar, lemon juice, shallot, garlic, mustard, salt, pepper, and thyme. Gradually add the oils. Keep the dressing refrigerated in a sealed container or bottle for up to one week. Shake well before serving.

# Mediterranean Grilled Chicken and Grape Skewers
( PICTURED ON PAGE 41 )

This is another good reminder that simple preparations can often be the best. These entrée skewers are easy to prepare ahead. They can also be threaded on shorter skewers and served as appetizers.

| MARINADE | ¼ | CUP PLUS 2 TABLESPOONS EXTRA-VIRGIN OLIVE OIL |
| | 2 | CLOVES GARLIC, MINCED |
| SERVES 4 | ½ | TEASPOON CRUSHED RED PEPPER FLAKES |
| | 1 | TABLESPOON CHOPPED FRESH OREGANO (1 TEASPOON DRIED) |
| | 1 | TABLESPOON CHOPPED FRESH ROSEMARY (1½ TEASPOONS DRIED) |
| | 1 | TEASPOON GRATED LEMON ZEST |
| | | |
| | 1 | POUND BONELESS, SKINLESS CHICKEN BREAST CUT INTO ¾-INCH DICE |
| | 1½ | CUPS GREEN GRAPES |
| | ½ | TEASPOON SALT |
| | 1 | TABLESPOON FRESH LEMON JUICE |

Make the marinade. In a small bowl, whisk together the ¼ cup olive oil, the garlic, red pepper flakes, oregano, rosemary, and lemon zest.

Thread alternating pieces of chicken and grapes onto 12-inch skewers. Place the skewers into a nonreactive dish or pan. Pour the prepared marinade over the skewers, turning to coat each one. Marinate in the refrigerator for 4 to 24 hours.

Remove the skewers from the marinade, and let excess oil drip off. Discard any remaining marinade. Season the skewers with the salt. Grill on a hot barbecue until the chicken is cooked through, 3 to 5 minutes on each side. Arrange on a serving platter and drizzle with the 2 tablespoons olive oil and the lemon juice.

# Spicy Chicken and Grape Lettuce Cups

Inspired by the cuisine of Thailand, each mouthful of this exotic salad is at once sweet, spicy, salty, and savory.

| | | |
|---|---|---|
| SAUCE | ¾ | CUP CHICKEN BROTH |
| | ¼ | CUP SOY SAUCE |
| | 3 TO 5 | TEASPOONS ASIAN CHILI PASTE WITH GARLIC |
| SERVES 4 | 1 | TABLESPOON CORNSTARCH |
| AS AN ENTRÉE | 1 | TABLESPOON WATER |
| OR 8 AS AN | 1 | TABLESPOON SUGAR |
| APPETIZER | 1 | TABLESPOON RED WINE VINEGAR |
| | | |
| | 1 | TABLESPOON PEANUT OR VEGETABLE OIL |
| | 6 | CLOVES GARLIC, MINCED |
| | 3 | TABLESPOONS MINCED FRESH GINGER |
| | 1½ | POUNDS GROUND CHICKEN BREAST |
| | 1½ | CUPS GREEN GRAPES, HALVED |
| | ¾ | CUP THINLY SLICED GREEN ONIONS, BOTH WHITE AND GREEN PARTS |
| | 1 | HEAD ICEBERG LETTUCE, CUT IN HALF AND SEPARATED INTO LEAVES |
| | 1 | SMALL CUCUMBER, PEELED AND CUT INTO JULIENNE STRIPS |
| | ¼ | CUP COARSELY CHOPPED CILANTRO |

Make the sauce. In a small bowl, combine the chicken broth, soy sauce, chili paste, cornstarch, water, sugar, and vinegar. Set aside.

Heat the oil in a large, nonstick skillet over high heat. Add the garlic and ginger and sauté for 30 seconds. Add the chicken and cook until it is no longer pink, 5 to 8 minutes. Add the grapes and green onions and sauté for 30 seconds more. Stir the sauce mixture and pour into the skillet. Cook, stirring constantly, until the sauce has thickened, 1 to 2 minutes. Remove from heat. Spoon the hot chicken mixture into the prepared lettuce leaves. Garnish with the cucumber and cilantro. Serve immediately.

# Grilled Seafood with Edamame and Green Grape Salad

Brightly flavored with a distinctive crunch, the Edamame and Green Grape Salad in this recipe can easily stand alone as a first course. But when you add chunks of juicy grilled seafood, it is transformed into a spectacular entrée. Substitute your favorite seafood such as halibut, salmon, mackerel, black cod, or calamari. Edamame (green soy beans) are now commonly available in the frozen vegetable sections of good supermarkets.

SERVES 4

| | |
|---|---|
| 1 | POUND ANY COMBINATION OF PLUMP OCEAN SCALLOPS AND PEELED, DEVEINED SHRIMP |
| 1½ | CUP LEMONGRASS VINAIGRETTE (SEE FACING PAGE) |
| 4 | HANDFULS (4 OUNCES) TENDER ASIAN GREENS SUCH AS TORN NAPA CABBAGE LEAVES, TATSOI, BABY SPINACH, WATERCRESS, AND/OR MIZUNA |
| 1 | CUP GREEN GRAPES |
| 1 | CUP DICED SEEDLESS CUCUMBER |
| 1 | CUP SHELLED EDAMAME |
| 1 | CUP PEELED, CHOPPED JICAMA |
| ⅓ | CUP LOOSELY PACKED FRESH BASIL LEAVES, TORN |
| ⅓ | CUP LOOSELY PACKED CILANTRO SPRIGS, TORN |

Marinate the seafood in ½ cup of the vinaigrette while the grill preheats and you assemble the salad ingredients.

Arrange 4 large plates or shallow bowls with a scattering of the Asian greens. In a large bowl, toss together the grapes, cucumber, edamame, jicama, basil, and cilantro.

Grill the scallops and shrimp on a preheated, hot grill for 2 minutes per side, or until just cooked through. Toss the grilled seafood with the grape mixture and moisten evenly with the remaining Lemongrass Vinaigrette. Spoon onto the greens and serve immediately.

| LEMONGRASS VINAIGRETTE | 1 | STALK LEMONGRASS, TENDER PART ONLY, SMASHED WITH THE SIDE OF A KNIFE AND CHOPPED |
| --- | --- | --- |
| | 2 | TABLESPOONS CHOPPED FRESH GINGER |
| MAKES 1½ CUPS | 1 | SMALL SHALLOT, CHOPPED |
| | 2 | CLOVES GARLIC, CHOPPED |
| | 1 | SMALL RED CHILE, SEEDED AND CHOPPED (OR ¼ TEASPOON RED PEPPER FLAKES) |
| | ¼ | CUP FRESH LIME JUICE |
| | ¼ | CUP RICE VINEGAR |
| | 1 | TABLESPOON SUGAR |
| | ¾ | CUP VEGETABLE OIL |
| | 1 | TEASPOON TOASTED SESAME OIL |
| | 1 | TEASPOON SALT |

In a blender, puree the lemongrass, ginger, shallot, garlic, chile, lime juice, rice vinegar, and sugar. Gradually add the vegetable oil and season with the sesame oil and salt. Store in the refrigerator and shake or stir well before serving.

# Seared Shrimp with Grapes, Citrus, and Rum

Family and friends will delight in this light, satisfying shrimp entrée accented with citrus, ginger, red grapes, and rum. It's especially tasty served over fluffy couscous to soak up the flavorful sauce.

SERVES 4

| | |
|---|---|
| 12 | LARGE SHRIMP, PEELED AND DEVEINED |
| ¼ | TEASPOON SALT |
| ¼ | TEASPOON PEPPER |
| 2 | TEASPOONS GRAPESEED OR VEGETABLE OIL |
| 1 | CUP RED GRAPES, SLICED |
| 1 | TABLESPOON SLIVERED LEMON ZEST |
| 1 | TABLESPOON SLIVERED LIME ZEST |
| 1 | JALAPEÑO PEPPER, SLICED INTO SLIVERS |
| 1 | TEASPOON SLIVERED GINGER |
| 3 | TABLESPOONS DARK RUM |
| 2 | TABLESPOONS BROWN SUGAR |
| 2 | TABLESPOONS FRESH LEMON JUICE |
| 1 | TABLESPOON FRESH LIME JUICE |

Pat the shrimp dry and season with the salt and pepper. Heat the oil in a large, nonstick skillet or sauté pan over medium-high heat. Sear the shrimp in the hot oil until they are bright pink and curling, about 2 minutes. Stir in the grapes, lemon zest, lime zest, jalapeño, and ginger. Pour on the rum, being careful to keep your hands and face away from any flame. Simmer for 1 minute. Stir in the brown sugar, lemon juice, and lime juice and simmer until the sauce is slightly syrupy and the shrimp are just cooked through, 2 to 3 minutes.

Arrange the shrimp on a serving platter or on individual dishes, spoon on the grape sauce, and serve.

# Riesling-and-Marmalade-Glazed Pork Chops
# with Fresh Grape Sauce

Thick, juicy pork chops are marinated in orange marmalade and spicy white Riesling wine. To cook, simply sear the chops in a pan and serve with a pan sauce of green grapes and orange zest.

| | | |
|---|---|---|
| SERVES 4 | ½ | CUP ORANGE MARMALADE |
| | ½ | CUP RIESLING OR GERMAN-STYLE WHITE WINE |
| | ½ | TEASPOON SALT |
| | 3 | WHOLE CLOVES |
| | | PINCH GROUND ALLSPICE |
| | FOUR | 6- TO 8-OUNCE THICK-CUT PORK CHOPS |
| | 1 | TABLESPOON VEGETABLE OIL |
| | 1 | CUP GREEN GRAPES, HALVED |
| | 2 | TABLESPOONS FINE STRIPS OF ORANGE ZEST |

In a medium bowl, whisk together the marmalade and Riesling. Add the salt, cloves, and allspice. Add the pork chops, then marinate in the refrigerator for at least 4 hours or overnight.

Heat the oil in a large skillet over medium-high heat. Lift the pork chops from the marinade, reserving the marinade. Sear the chops in the hot oil until they are brown, about 3 minutes per side. The marinated chops will brown quickly, so take care that they do not burn.

Pour in half the remaining marinade and bring to a boil. Reduce the heat to low and simmer, covered, until the pork chops are cooked through, about 6 minutes more. Check the pan and turn the chops often so they do not scorch. When the chops are cooked through, move them to a platter and keep warm. Cook the remaining sauce until it is a nice, syrupy gravy. Add the grapes and orange zest. Stir well and then spoon over the pork chops. Serve immediately.

# Seared Duck Breast with Black Grapes and Port Reduction

Three main ingredients are all it takes to make this simply spectacular entrée. It is the perfect romantic meal for two, and an ideal Valentine's Day or anniversary surprise. Serve with roasted potatoes and steamed asparagus or green beans. Plump, lean duck breasts are becoming more common in fine markets. This recipe can also be adapted with great success using seared calf's liver or veal scallops.

SERVES 2

2 CUPS PORT WINE
¾ CUP BLACK GRAPES, HALVED
2 PLUMP, LEAN DUCK BREASTS
SALT
PEPPER

In a medium saucepan over high heat, boil the port until it has reduced to ½ cup. Stir in the grapes and set aside. Preheat the oven to 425 degrees F. Season the duck breasts with salt and pepper. Heat an ovenproof skillet over medium-high heat and place the duck, skin side down, in the pan. Cook the duck breasts until they are a nice crispy brown, about 4 minutes. Turn the duck breasts over, brown the other side, then place the skillet in the preheated oven and cook for 4 to 6 minutes. Foodies prefer duck breast served medium-rare, but if you prefer, cook the duck until the internal temperature reaches 160 degrees F. Like all meats, allow the duck breasts to rest at least 2 to 3 minutes before slicing. Slice and fan the duck onto serving plates and spoon on the warm reduced port and grapes. Serve immediately.

# Beef Brisket Braised with Grapes and Coriander

This slow-cooked brisket has the aroma and depth of flavor of meats that are cooked with wine, but it is alcohol-free. The meat is so succulent it will melt in your mouth. The cooking juices can be reduced down to make a nice rich gravy to serve alongside creamy polenta or buttered noodles.

SERVES 6

| | |
|---|---|
| 1 | BEEF BRISKET (4 TO 5 POUNDS) |
| 1 | TABLESPOON VEGETABLE OIL |
| 1 | YELLOW ONION, CHOPPED (ABOUT 2 CUPS) |
| 2 | CLOVES GARLIC, CHOPPED |
| 1½ | CUPS RED OR BLACK GRAPES |
| 3 | TABLESPOONS ALL-PURPOSE FLOUR |
| 2 | TABLESPOONS GROUND CORIANDER |
| 2 | CUPS GOOD-QUALITY BEEF STOCK |
| 1 | TEASPOON SALT, PLUS MORE TO TASTE |
| ½ | TEASPOON PEPPER, PLUS MORE TO TASTE |
| | SPRIG OF THYME |
| 1 | BAY LEAF |
| 2 | TABLESPOONS CHOPPED FRESH PARSLEY FOR GARNISH |

Preheat oven to 325 degrees F. Trim excess fat from the beef brisket. Heat the oil in a large, heavy-lidded casserole over medium-high heat. Sear the brisket to a dark brown, 4 to 6 minutes per side. Remove the meat to a plate and set aside. If the residue on the bottom of the pan seems not just dark brown but scorched or burnt, wash out the pan, add more oil, and then proceed. Sauté the onion and garlic in the oil until golden brown, about 4 minutes. Stir in the grapes, flour, and coriander. Place the brisket, fatty side up, on top of the base of grapes and onion. Add the stock, salt, pepper, thyme, and bay leaf. Bring the liquids just to a boil, cover, and place in the preheated oven to cook for 2 hours, or until the brisket is very tender.

Lift the brisket from the cooking juices onto a cutting board and slice the meat against the grain. To make a gravy, strain the cooking juices into a saucepan and boil until the gravy has reduced to a slightly syrupy consistency, about 5 minutes. Add additional salt and pepper if needed.

Serve the brisket on a platter, moistened with some of the gravy and garnished with chopped parsley. Serve remaining gravy alongside.

# Pantry Stuffed French Toast

( PICTURED ON PAGE 38 )

This recipe is very adaptable, ideally made with some of the leftover ingredients
from the previous evening's gourmet meal. Here, baguette slices are filled with almond
paste, red grapes, and Neufchâtel cheese to make a very refined and elegant brunch dish.
Try big slices of chewy rosemary or potato bread and substitute some ricotta and a bit of
blue or Parmesan cheese for a bold and hearty autumn or winter breakfast. Or, try challah,
potato loaf, or rosemary bread with a combination of well-drained ricotta, mascarpone,
a chunk of Brie, triple-cream Saint André, or even a bit of mild blue cheese.

SERVES 4 TO 6

| | |
|---|---|
| 1 | BAGUETTE |
| ¼ | CUP ALMOND PASTE |
| ONE | 8-OUNCE BLOCK NEUFCHÂTEL OR CREAM CHEESE |
| 1 | CUP RED GRAPES, QUARTERED OR CHOPPED |
| 3 | EGGS |
| 1½ | CUPS MILK |
| 1 | TEASPOON SUGAR |
| | PINCH OF SALT |

ORANGE SYRUP

| | |
|---|---|
| ½ | CUP ORANGE MARMALADE |
| 1 | CUP WATER |

| | |
|---|---|
| 2 | TEASPOONS BUTTER OR VEGETABLE OIL FOR FRYING |
| ¼ | CUP SLIVERED ALMONDS FOR GARNISH (OPTIONAL) |

Slice the bread, only cutting through the bottom crust every other cut so you form a "pocket" of two thin slices of bread connected at the base with the crust.

With a mixer, blend the almond paste with ¼ cup of the cheese until smooth. Add the remainder of the cheese and mix well. Fold in the grapes. Spoon a heaping tablespoon of the filling into the prepared bread pockets, smooth to the edges, and press lightly to flatten. Continue with the remaining bread and filling.

Whisk together the eggs, milk, sugar, and salt in a pie plate or shallow dish. Make the syrup by melting together the marmalade and the water in a small saucepan over medium heat.

Heat a griddle to medium and lightly grease with the butter or oil. Dip the prepared bread into the egg mixture and turn to coat lightly. Fry the French toast slowly, about 4 minutes per side, until golden brown on the outside and just warmed through. Keep finished toast warm in a 200-degree-F oven while cooking the rest. Serve warm with the syrup and, if desired, a sprinkling of almonds.

Sparkling Grape
and Berry Jells

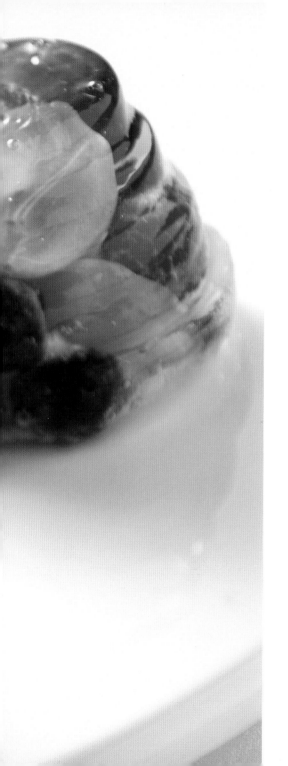

:::::: classics

Grapes are a classical food. Their beauty is depicted in ancient sculpture and Old Master paintings. They appear in the Bible, the Talmud, and the world's oldest cookbooks. California's grape tradition began with the Spanish missions, with grapes planted for sacramental wine. In the early 1800s, Californians began growing grapes to eat fresh. In the 1900s, European immigrants came to California's great Central Valley, planting vines and working the land to produce fresh grapes for a growing nation. Today, the descendants of those early pioneers are still farming today. Its unique climate is a primary reason why more than 98 percent of the fresh grapes grown in the United States come from California.

Sole Véronique

# Sole Véronique

( PICTURED ON PAGE 59 )

Tender sole fillets are gently poached and napped with a delicate cream sauce dotted with cold green grapes. It's a French classic that is overdue for a comeback.

| SERVES 6 | | |
|---|---|---|
| | 6 | LARGE OR 12 SMALL FRESH SOLE FILLETS (ABOUT 1½ POUNDS) |
| | ½ | CUP DRY WHITE WINE |
| | 1 | SMALL ONION SLICE |
| | 2 TO 3 | SLICES FRESH LEMON |
| | | SPRIG OF THYME |
| | 5 TO 6 | PEPPERCORNS |
| | 1 | TABLESPOON BUTTER |
| | 1 | TABLESPOON FLOUR |
| | ½ | CUP HALF-AND-HALF |
| | ½ | TEASPOON SALT |
| | ¼ | TEASPOON WHITE PEPPER |
| | 1 | CUP GREEN GRAPES |
| | ½ | TEASPOON GRATED LEMON ZEST |

Preheat oven to 325 degrees F. Lightly butter a glass or ceramic ovenproof baking dish.

Fold the sole fillets in half, skin side in. Place the fillets in the prepared dish with the wine, onion, lemon slices, thyme, and peppercorns. Cover the dish loosely with foil and bake for 10 to 12 minutes for small fillets and 16 to 18 minutes for larger fillets, or until the fish is just cooked through. Carefully lift the fish from the pan and place on a warm serving dish. Strain any cooking liquids and reserve. Discard the solids. To make the sauce, melt the butter in a medium saucepan over medium heat. Whisk in the flour and cook until the roux is bubbly and golden brown. Remove the pan from the heat and pour in the strained cooking liquids. Stir well. Return the pan to the heat, add the half-and-half, and simmer the sauce for 3 minutes, stirring constantly. Season with the salt and white pepper. Add the grapes and lemon zest. Spoon the sauce over the sole and serve immediately.

# Pâté and Waldorf Salad Baguette Sandwiches

These marvelous sandwiches are an entire picnic rolled into a single dish. You can assemble and serve them immediately or wrap them in paper and take them to a picturesque location. Making the filling in advance helps to marry the ingredients into a more cohesive and flavorful salad. Spanish Marcona almonds are available at good supermarkets and specialty stores.

SERVES 4

1 THICK SLICE (4 OUNCES) COUNTRY-STYLE PÂTÉ, DICED
½ CUP GRATED GRUYÈRE CHEESE
¼ CUP MAYONNAISE
1 TABLESPOON DIJON MUSTARD
½ CUP GREEN, RED, AND/OR BLACK GRAPES, QUARTERED OR SLICED
¼ CUP GOOD TOASTED AND SALTED ALMONDS, SUCH AS MARCONA ALMONDS
1 SMALL CELERY STALK, DICED
½ MEDIUM TART APPLE, DICED
1 VERY FRESH BAGUETTE
LETTUCE LEAVES AND TOMATO SLICES (OPTIONAL)

Mash together the chopped pâté and grated cheese. Add the mayonnaise and Dijon mustard and mix well. Stir in the grapes, almonds, celery, and apple. Chill until ready to use.

Slice the baguette lengthwise without cutting completely through. Press down the middle of the bread to make a shallow channel. Spoon the salad evenly into the channel. (Add lettuce and tomato if desired.) Cut the prepared sandwich into 4 even portions and serve. Wrap the sandwiches in paper if transporting.

# Curried Chicken Salad

Having a batch of Curried Chicken Salad on hand means you are prepared for almost any occasion. Serve a scoop of this classic on a bed of greens for a light dinner, fill a pita, or simply carry a container to work. You might even go crazy now and then and make the salad with bay shrimp.

MAKES 4 CUPS

12 OUNCES COOKED CHICKEN MEAT, DICED (3 CUPS)

1 MEDIUM RED APPLE, DICED (1 CUP)

¾ CUP GREEN, RED, AND/OR BLACK GRAPES, HALVED

1 MEDIUM CELERY STALK, DICED

¼ CUP (1 OUNCE) BLANCHED SLIVERED ALMONDS (OPTIONAL)

½ CUP MAYONNAISE, OR PLAIN YOGURT, OR ANY COMBINATION OF THE TWO

2 TABLESPOONS MANGO CHUTNEY

1 TABLESPOON CURRY POWDER

½ TEASPOON SALT

½ TEASPOON PEPPER

FRESH CILANTRO SPRIGS AND SLICED GREEN ONIONS FOR GARNISH (OPTIONAL)

In a large bowl, combine the chicken, apple, grapes, celery, and almonds, if using. Add the mayonnaise, chutney, and curry powder and stir to mix and evenly moisten. Season with the salt and pepper. Keep refrigerated until ready to use. Serve garnished with cilantro and green onions, if desired.

# Sparkling Grape and Berry Jells

( PICTURED ON PAGE 56 )

It would just be wrong to not include some kind of dish with grapes suspended in gelatin. This is a much more sophisticated and elegant recipe than many of the church supper staples. It is not too sweet and has a lively effervescence. It can be made in small individual molds or in one large mold.

SERVES 8

| | |
|---|---|
| ¾ | CUP WATER |
| 3 | (¼ OUNCE) ENVELOPES PLAIN, UNFLAVORED GELATIN |
| 2 | TABLESPOONS SUGAR |
| 1½ | CUPS GREEN OR RED GRAPES, HALVED |
| 2 | CUPS ASSORTED FRESH BERRIES SUCH AS RASPBERRIES, BLACKBERRIES AND BLUEBERRIES |
| 1 | BOTTLE (24.5 OUNCES) CHILLED SPARKLING WHITE GRAPE JUICE |

Put the water in a small saucepan and sprinkle the gelatin over the top. Leave to soften for 1 minute. Place the pan over low heat and gently melt, stirring often. Add the sugar and continue to cook for about 1 minute, until the sugar is dissolved. Remove the pan from the heat and let cool to room temperature.

Meanwhile, prepare 1 8-cup or 8 1-cup molds by filling them with a mixture of grapes and berries. Pour the sparkling grape juice into a bowl and add the cooled gelatin mixture. Pour the juice and gelatin mixture into the prepared molds. Refrigerate until firm, at least 2 hours or overnight.

To loosen the gelatin molds, quickly dip the molds in a pan filled with warm water. Invert and serve.

## GRAPES WITH BROWN SUGAR AND SOUR CREAM
## (AKA GRAPES ROMANOV)

This is a classic American dessert and a favorite of the late, great James Beard. Simply arrange halved grapes in individual serving dishes. Sprinkle heavily with brown sugar and top with sour cream. Chill until ready to serve.  O  A slightly fancier alternative is sometimes called Grapes Romanov. Arrange halved grapes in individual heatproof serving dishes as before, but top first with sour cream and then brown sugar. Chill for at least 1 hour so the brown sugar melts into the sour cream, then caramelize or "brûlée" the sugar with a small torch or by placing the dishes under a preheated broiler until the sugar is bubbly, about 2 minutes. Serve immediately.

# Margaret's Special Grape Pie

California's fresh grape farmers enjoy their product as much as anyone. The tradition of family farming brings a strong sense of stewardship to their work as each generation learns a respect for the land, for the art of growing fresh grapes, and for the traditions handed down through the years. This traditional recipe comes from Margaret Zaninovich, whose family has been growing fresh grapes in California for over 100 years.

SERVES 8

PASTRY FOR A DOUBLE-CRUST 9-INCH PIE (PAGE 68)

5 CUPS BLACK GRAPES, HALVED
½ CUP SUGAR
4 TABLESPOONS CORNSTARCH
½ TEASPOON GROUND CINNAMON
½ TEASPOON SALT
1 TABLESPOON FRESH LEMON JUICE
2 TABLESPOONS BUTTER

Preheat oven to 425 degrees F.

Line a 9-inch pie plate with pastry. In a medium bowl, stir together the grapes, sugar, cornstarch, cinnamon, and salt. Pour into the pastry. Sprinkle with the lemon juice and dot with the butter. Top with the remaining pastry and press to seal. Flute the edges and cut vents into the top of the pie. Bake in the preheated oven for 10 minutes, then reduce the heat to 350 degrees F and cook 30 to 35 minutes more, or until the filling in the center is bubbly.   >>>

Margaret's Special Grape Pie
( CONTINUED )

| BASIC | 3 | CUPS ALL-PURPOSE FLOUR |
| PIE PASTRY | 1 | TABLESPOON SUGAR |
| MAKES | ½ | TEASPOON SALT |
| ENOUGH FOR A | ¾ | CUP COLD VEGETABLE SHORTENING |
| DOUBLE-CRUST | ¼ | CUP COLD BUTTER |
| 9-INCH PIE | 6 TO 8 | TABLESPOONS ICE WATER |

In a large bowl combine the flour, sugar, and salt. Using the largest holes on a box grater, grate the shortening and butter into the flour mixture. Sift the mixture through your fingers until it is crumbly. Stir in the cold water, adding just enough so the pastry comes together in a ball. Divide the dough into two disks, wrap with plastic, and chill until ready to use.

## SUGARED AND SPICED GRAPES ON THE VINE

Fresh grapes shimmering with a crystallized sugar crust is a classic garnish for desserts and cheese platters. And while it does, in some ways, feel like gilding a lily, we are still charmed by the idea and have even expanded it to include an assortment of additional toppings. Grind up your favorite topping to a fine dust in a coffee or spice mill.

Whisk an egg white until soft peaks form. Brush clusters of grapes evenly with the frothy egg white and sprinkle with your choice of topping. Set the dusted/coated grapes onto a fine-mesh cooling rack to dry thoroughly. Refrigerate until ready to use.

WHITE SUGAR
FINELY GROUND, ROASTED NUTS SUCH AS PISTACHIOS,
   HAZELNUTS, ALMONDS, PINE NUTS, OR PEANUTS
FINELY GROUND, ROASTED ALMONDS OR PEANUTS
   WITH A PINCH OF CURRY POWDER
FINELY GROUND, ROASTED ALMONDS OR PEANUTS
   WITH A PINCH OF COCOA POWDER AND CAYENNE PEPPER
GROUND TOASTED SESAME SEEDS
FINELY GROUND BISCOTTI
VERY FINELY GRATED PARMESAN CHEESE
GROUND TOASTED LIGHTLY-SWEETENED COCONUT

Grilled Onion and
Grape Relish

:::::: sides

Americans consume more than
8 pounds of fresh grapes each year.
Their simple portability makes them
an ideal fit for today's active world.
At home or on the go, fresh grapes are
a quick and healthful snack. They also
speed up side dishes for family meals.
A flash of flavor in a pasta, a sweet note
in a slaw, or a surprise twist to salsa—fresh
grapes are an ideal addition to any meal.

Fruited Tabbouleh

# Grilled Onion and Grape Relish

( PICTURED ON PAGE 70 )

This is especially tasty made with the sweet onions of early summer. Use as a topping for grilled fish or chicken.

SERVES 4 TO 6

| | |
|---|---|
| 1 | SWEET ONION, CUT INTO ¼-INCH SLICES |
| 1 | CUP RED GRAPES, HALVED |
| ¼ | CUP CHOPPED FRESH MINT |
| 3 | TABLESPOONS FRESH LEMON JUICE |
| 2 | TABLESPOONS EXTRA-VIRGIN OLIVE OIL |
| ¼ | TEASPOON SALT |
| ¼ | TEASPOON PEPPER |

Grill the onion slices on a preheated, medium-hot grill for 3 minutes per side, or until soft and beginning to brown. Put the onions into a medium-sized bowl and stir in the grapes, mint, lemon juice, olive oil, salt, and pepper and toss well.

# Green Grape and Tomatillo Salsa

Serve with chips for dipping. This is also great with wedges of warm cheese quesadilla, or grilled chicken or fish.

| MAKES | 6 TO 7 | MEDIUM TOMATILLOS (8 OUNCES), HUSKED AND RINSED |
|---|---|---|
| 2 CUPS | 1¾ | CUPS GREEN GRAPES |
| | 1 | JALAPEÑO PEPPER, SEEDED AND CHOPPED |
| | 2 | CLOVES GARLIC, CHOPPED |
| | 2 | TABLESPOONS CHOPPED FRESH CILANTRO |
| | 1 | TABLESPOON FRESH LIME JUICE |
| | ½ | TEASPOON SALT |

Blend all the ingredients in a food processor, pulsing until uniformly coarse in texture. Or, mince tomatillos, grapes, jalepeño, garlic, and cilantro and mix together in a bowl. Stir in the lime juice and salt. Keep refrigerated until ready to serve.

# Spicy Carrot Slaw

Peanuts and cilantro add a fresh and contemporary twist to that ancient carrot-and-raisin salad of your childhood. This slaw is crisp and crunchy, with a tangy orange and chili zip.

SERVES 8

- ⅓ CUP ORANGE JUICE CONCENTRATE, UNDILUTED
- 2 TEASPOONS FRESH LIME JUICE
- 1 SMALL GREEN CHILE, SEEDED AND MINCED (USE A JALAPEÑO, SERRANO, OR EVEN HABANERO IF YOU LIKE IT REALLY HOT!)
- ½ TEASPOON SALT
- 3 LARGE PEELED CARROTS, SHREDDED
- 2 CUPS GREEN, RED, AND/OR BLACK SEEDLESS GRAPES, HALVED
- ½ MEDIUM JICAMA, PEELED AND JULIENNED (1½ CUPS)
- ½ CUP CHOPPED FRESH CILANTRO
- ½ CUP SALTED PEANUTS, COARSELY CHOPPED

In a small bowl, whisk together the orange juice concentrate, lime juice, chile, and salt. In a separate serving bowl, combine the carrots, grapes, jicama, cilantro, and peanuts. Pour the dressing over the salad and toss to blend. Keep refrigerated until ready to serve.

# Fruited Tabbouleh

( PICTURED ON PAGE 73 )

Fresh grapes and orange take the place of tomato and masses of garlic in this variation of a classic bulgur wheat salad. This seems to suit a wider audience while still retaining the salad's healthy appeal.

SERVES 6

| | |
|---|---|
| 2½ | CUPS CHICKEN OR VEGETABLE BROTH |
| 1½ | CUPS BULGUR WHEAT |
| 1½ | CUPS GREEN, RED, AND/OR BLACK GRAPES |
| 1 | MEDIUM ORANGE, PEELED AND DICED |
| ½ | CUP MINCED RED ONION |
| ¼ | CUP CHOPPED FRESH MINT LEAVES |
| ⅓ | CUP EXTRA-VIRGIN OLIVE OIL |
| 3 | TABLESPOONS FRESH LEMON JUICE |
| 1 | TABLESPOON SUGAR |
| ¾ | TEASPOON SALT |
| ¾ | TEASPOON GROUND GINGER |
| ¾ | TEASPOON GROUND CUMIN |
| ½ | TEASPOON PEPPER |

In a medium saucepan, bring the broth to a boil. Stir in the bulgur, cover, and remove from the heat. Let rest for 25 minutes. Fluff with a fork and cool completely.

In a large bowl, stir together the bulgur, grapes, orange, onion, and mint. In a small bowl, whisk together the oil, lemon juice, sugar, salt, ginger, cumin, and pepper. Pour the dressing over the bulgur mixture and toss well. Keep chilled until ready to serve.

# Sautéed Sweet Potatoes with Bacon, Kale, and Grapes

Sweet potatoes and grapes are another wonderful flavor combination. This makes an especially good side dish for Thanksgiving and Christmas. For a variation, substitute cubes of tender winter squash for the sweet potatoes.

| SERVES 6 | | |
|---|---|---|
| | 3 | SLICES BACON, CHOPPED |
| | ½ | MEDIUM ONION, SLICED |
| | 2 | CUPS LIGHTLY PACKED STEMMED, TORN KALE |
| | ½ | CUP FRESH ORANGE JUICE |
| | ¼ | TEASPOON SALT |
| | ¼ | TEASPOON PEPPER |
| | 2 | MEDIUM SWEET POTATOES OR YAMS, COOKED, PEELED, AND CUT INTO ½-INCH DICE |
| | 1 | CUP GREEN, RED, AND/OR BLACK GRAPES, HALVED |

In a large covered skillet or sauté pan, cook the bacon over medium-high heat until it is just brown. Add the onion and cook until it is golden and aromatic, about 3 minutes. Add the kale, orange juice, salt, and pepper. Stir well and cover. Cook on medium heat, stirring occasionally, until the kale is tender, 3 to 4 minutes. Carefully stir in the cooked sweet potatoes and grapes. Cover and cook 2 minutes more, until the potatoes are warmed through and the grapes are just soft. Stir gently to mix and serve warm.

# Creamy Penne with Blue Cheese and Grapes

Blue cheese and grapes are a classic combination. Now imagine how good this pair is blended into creamy pasta and served alongside a peppery, grilled porterhouse steak, a crisp roasted chicken, or a fillet of fresh salmon. But there is no need to wait for company. Go ahead and pamper yourself with a bowl of this upscale comfort food.

SERVES 4 TO 6

| | |
|---|---|
| 8 | OUNCES PENNE PASTA |
| 1 | TABLESPOON BUTTER |
| 1 | CLOVE GARLIC, MINCED |
| 1 | CUP HEAVY CREAM |
| ¾ | CUP GREEN, RED, AND/OR BLACK GRAPES, HALVED |
| ½ | CUP BLUE CHEESE, CRUMBLED |
| ¼ | TEASPOON SALT |
| ¼ | TEASPOON PEPPER |

Cook the pasta according to the manufacturer's directions in plenty of boiling water until it is just al dente. Drain and set aside.

Melt the butter in a large skillet or sauté pan over medium-high heat. Add the garlic and sauté until tender and aromatic but not brown, about 1 minute. Pour in the cream and bring to a boil. Cook on high until the cream has thickened slightly, 4 to 5 minutes. Add the pasta, grapes, and blue cheese and toss to coat. Season with the salt and pepper and serve hot.

Mediterranean Grape Pitas

:::::::: quick

Quick. Simple. No fuss. Fill the plate.
Enhance an entrée. Delight your guests.
Feed your family. Surprise yourself.
Enjoy your food. Give yourself
time away from the stove.
Turn a simple weekday meal
into a special-occasion delight
in just a few minutes.

Ham Steak with
Quick Grape Pan Sauce

# Ham Steak with Quick Grape Pan Sauce
( PICTURED ON PAGE 85 )

A good bone-in ham steak makes a quick family meal. Grapes and mango chutney make a sweet and spicy pan sauce. Serve with cheese grits or scalloped potatoes and a crispy green salad. If the mango chutney has been made with large pieces, be sure and chop them up before adding.

SERVES 4

| | |
|---|---|
| ONE | 1-POUND HAM STEAK |
| ½ | CUP WHITE WINE, OR ORANGE JUICE |
| 3 | TABLESPOONS MANGO CHUTNEY |
| ¾ | CUP GREEN, RED, AND/OR BLACK GRAPES, HALVED |
| 2 | TABLESPOONS COLD BUTTER, CUT INTO 6 PIECES |

Brown the ham steak in a preheated nonstick skillet over medium-high heat, 2 to 3 minutes per side. Remove from the heat and arrange the ham on a warmed platter or individual plates. Pour the white wine into the skillet, being careful to protect your hands and face from any steam. Stir in the mango chutney and grapes and return to the stove top. Simmer over medium heat, stirring constantly, until the chutney is melted and the grapes are just warmed through, about 1 minute. Drop the pieces of cold butter into the pan and whisk in until just melted. Pour the pan sauce over the ham steak and serve.

# Sweet and Spicy Moroccan Turkey Pilaf

It's quick, lean, and appealing to all ages. What more could you want from a midweek meal? If you prefer to cook the pilaf in the oven, follow the instructions as written, but instead of covering and leaving to cook on the burner, simply pop the dish in a preheated 325 degree F oven for 30 to 40 minutes.

SERVES 4 TO 6

| | |
|---|---|
| 2 | TEASPOONS VEGETABLE OIL |
| 1 | POUND SKINLESS, BONELESS TURKEY BREAST, CUT INTO STRIPS |
| 1 | CUP LONG GRAIN RICE |
| 1½ | CUPS CHICKEN BROTH |
| 1 | TEASPOON SALT |
| ½ | TEASPOON PEPPER |
| ½ | TEASPOON GROUND CINNAMON |
| ¼ | TEASPOON CAYENNE PEPPER |
| 1 | CUP GREEN, RED, AND/OR BLACK GRAPES |
| ¼ | CUP PINE NUTS OR SLIVERED ALMONDS |
| 6 TO 7 | (¼ CUP) DRIED APRICOTS, SLICED |
| 2 | GREEN ONIONS, BOTH WHITE AND GREEN PARTS, SLICED |
| 2 | TABLESPOONS CHOPPED FRESH MINT |

In a large covered skillet, heat the oil over medium-high heat. Brown the turkey on all sides. Stir in the rice and cook for 1 minute. Pour in the chicken broth and season with the salt, pepper, cinnamon, and cayenne. Stir until the mixture comes to a boil, then cover and reduce the heat to a simmer. Cook for 30 minutes, or until the rice is tender. Remove the pilaf from the heat and gently stir in the grapes, pine nuts, apricots, green onions, and mint. Serve warm.

# Slow-Cooked Pork Chops with Spiced Honey and Grapes

This dish is quick to assemble. The savory flavor and marvelous cooking aromas make it particularly comforting on brisk days. Serve with mashed potatoes or a cheesy baked pasta to soak up the rich, spiced gravy. It is especially convenient when made in a slow cooker.

SERVES 6

| | |
|---|---|
| 2 | TABLESPOONS BUTTER OR MARGARINE |
| 3 | POUNDS THICK-CUT PORK CHOPS OR BONELESS, COUNTRY-STYLE SPARERIBS |
| 4 | TABLESPOONS FLOUR |
| 1½ | CUPS RED OR BLACK GRAPES |
| ⅓ | CUP HONEY |
| 2 | CINNAMON STICKS, BROKEN IN HALF |
| 3 TO 5 | WHOLE CLOVES |
| ½ | TEASPOON GROUND GINGER |
| ½ | TEASPOON SALT |
| ¼ | TEASPOON PEPPER |

Preheat oven to 325 degrees F.

Melt the butter in a heavy, lidded sauté pan or casserole over medium-high heat. Dust the pork with the flour and brown slowly in the melted butter. Cook in batches if necessary.

Remove the pan from the heat to prevent the flour from scorching and stir in the grapes, honey, cinnamon, cloves, ginger, salt, and pepper. Cover and bake in the preheated oven for 50 to 60 minutes, or until the pork is tender. Serve hot.

Slow-cooker variation:  Brown the pork on the stove top and then put it in the slow cooker with the remaining ingredients. Cook on low for 6 to 8 hours (or according to the manufacturer's instructions) and serve hot.

# Pizza with Grapes and Fontina

You think you have nothing in the house to make for dinner? Look again. A good homemade pizza can be tossed together in minutes. Fontina cheese is great to keep on hand as an alternative to mozzarella. Of course, Parmesan is also very good here.

| SERVES 2 | ONE | 14-INCH PIZZA CRUST (RECIPE FOLLOWS) |
|---|---|---|
| | 1 | TABLESPOON EXTRA-VIRGIN OLIVE OIL |
| | 1 | SMALL CLOVE GARLIC, MINCED |
| | ¼ | TEASPOON DRIED OREGANO |
| | | PINCH OF RED PEPPER FLAKES |
| | 1 | CUP GRATED FONTINA CHEESE |
| | ¾ | CUP RED GRAPES, HALVED |

Preheat oven to 500 degrees F. Arrange the pizza crust on a baking stone or pizza pan. Drizzle the olive oil on the pizza and sprinkle evenly with the garlic, oregano, and red pepper flakes. Top with the fontina and grapes. Bake the pizza in the preheated oven for 12 to 15 minutes, or until the crust has browned and the cheese is bubbly.

Slice into wedges and serve.

| PIZZA DOUGH | 1½ | CUPS WARM WATER |
|---|---|---|
| | ONE | ¼-OUNCE ENVELOPE (¾ TABLESPOON) DRY YEAST |
| MAKES TWO | 1 | TEASPOON SUGAR |
| 14-INCH THIN | 1 | TEASPOON SALT |
| CRUST PIZZAS | 4 | CUPS ALL-PURPOSE FLOUR |
| | 3 | TABLESPOONS EXTRA-VIRGIN OLIVE OIL |

In a small bowl, stir together the water, yeast, and sugar. Let the mixture rest until the yeast blooms, 3 to 4 minutes.    >>>

Place the salt and three cups of the flour in a mixing bowl. Add the yeast mixture and the oil, and stir together to form a sticky dough. Knead, adding the remaining flour a little at a time, until the dough becomes a smooth, elastic ball, 5 to 8 minutes. If you prefer to use an electric mixer, combine the ingredients with the paddle attachment and then change to the dough hook. Add the flour a little at a time until the dough is smooth and comes away from the sides of the bowl, 2 to 4 minutes.

Turn the dough out into a lightly oiled bowl and cover with a dry cloth or plastic wrap. Let rise for 30 minutes before stretching or rolling into the desired shape.

## Breakfast Trifles

One way to get you or your family to eat a healthy breakfast is to make foods that look more like dessert! If you make these trifles in clear plastic cups, they are even easier to grab as you're running out the door.

| SERVES 6 | 2 | LARGE OR 3 MEDIUM BRAN MUFFINS (DAY-OLD IS FINE) |
|---|---|---|
| | 1/3 | CUP ORANGE JUICE |
| | 2½ | CUPS (1 POUND) GREEN, RED, AND/OR BLACK GRAPES |
| | 3 | CUPS PLAIN YOGURT |

Cut the muffins into 1-inch cubes. In six 10-ounce glasses, layer the muffin pieces, a few drops of orange juice, grapes, and yogurt. Repeat with another layer. Cover with plastic wrap and chill overnight, or until ready to serve. These will hold in the refrigerator for up to 3 days.

# Smoked Turkey Salad with Cashews and Sherry Dressing

This is a big-time favorite for anyone who tries it. Substitute roasted turkey or chicken for smoked turkey if you prefer.

SERVES 6

| | |
|---|---|
| 2 | CUPS (8 TO 10 OUNCES) DICED SMOKED TURKEY MEAT |
| 1 | CUP SEEDLESS GRAPES |
| 1 | CELERY STALK, SLICED |
| ¼ | CUP ROASTED, SALTED CASHEW PIECES, PLUS MORE FOR GARNISH (OPTIONAL) |
| 1 | GREEN ONION, BOTH WHITE AND GREEN PARTS, SLICED |
| 2 | TABLESPOONS CHOPPED FRESH PARSLEY |
| ⅓ | CUP MAYONNAISE |
| 2 | TABLESPOONS MEDIUM DRY SHERRY |
| ½ | TABLESPOON FRESH LEMON JUICE |
| ½ | TEASPOON SUGAR |
| ¼ | TEASPOON SALT |
| ¼ | TEASPOON PEPPER |

In a large bowl, mix together the turkey, grapes, celery, cashews, onion, and parsley. Add the mayonnaise, sherry, lemon juice, sugar, salt, and pepper and mix until evenly moistened. Chill until ready to serve. Garnish with additional cashews if desired.

# Italian Sausages with Grapes and Basil

Sweet or spicy Italian sausages work with this. In fact, almost any sausage works. Try bratwurst, kielbasa, potato, or chicken and apple sausages. For dinners on the run, fill a hoagie roll or hot dog bun with a sausage and top with the juicy onions and grape sauce.

SERVES 6

| | |
|---|---|
| 1 | TABLESPOON VEGETABLE OIL |
| 6 | ITALIAN SAUSAGE LINKS |
| ½ | MEDIUM ONION, SLICED |
| 1½ | CUPS RED OR BLACK GRAPES |
| 2 | TABLESPOONS BALSAMIC VINEGAR |
| ¼ | CUP SLICED FRESH BASIL |

Heat the oil in a large, covered skillet or sauté pan over medium-high heat. Prick the sausage skins all over with a fork to prevent them from bursting. Sear the sausages in the oil, turning often, until they are an even brown, about 8 minutes. (They may not be completely cooked through at this point.) Remove the sausages from the pan and set aside.

Sauté the onion in the pan until it is brown and slightly tender, about 3 minutes. Add the grapes and balsamic vinegar and stir to mix. Place the sausages on top of this mixture, cover, and reduce the heat to medium-low. Continue cooking until the sausages are cooked through, about 4 minutes or until the internal temperature reaches 160 degrees F. Remove the cover and let the sauce reduce slightly, about 2 minutes more. Sprinkle the dish with plenty of fresh basil and serve warm.

# Mediterranean Grape Pitas
( PICTURED ON PAGE 82 )

These are especially good made with thin slices of cold rare lamb, but they work well with most any leftover roasted meat.

| | | |
|---|---|---|
| TAHINI DRESSING | 1 | CUP PLAIN YOGURT |
| | 2 | TABLESPOONS TAHINI (GROUND SESAME SEED PASTE) |
| | 1 | CLOVE GARLIC, MINCED |
| | 1/4 | TEASPOON GROUND CUMIN |
| | | PINCH OF SALT |
| | | DASH OF TABASCO SAUCE (OPTIONAL) |

| | | |
|---|---|---|
| MAKES 6 PITA SANDWICHES | 1 | CUP CANNED GARBANZO BEANS, RINSED AND DRAINED |
| | 1 | CUP GREEN, RED, AND/OR BLACK GRAPES, HALVED |
| | 3/4 | CUP CRUMBLED FETA CHEESE |
| | 1/3 | CUP DICED RED ONION |
| | 2 | TABLESPOONS CHOPPED FRESH MINT |
| | 1/2 | TEASPOON DRIED OREGANO |
| | | SALT |
| | | FRESHLY GROUND BLACK PEPPER |
| | 6 | GREEN LETTUCE LEAVES, WASHED AND DRIED |
| | 6 | PIECES OF PITA OR SOFT FLAT BREAD |
| | 1 | POUND THINLY SLICED COLD, COOKED LAMB, OR VIRTUALLY ANY, SLICED COLD ROAST OR COOKED SEAFOOD |

Make the tahini dressing by combining the yogurt, tahini, garlic, cumin, salt, and Tabasco (if desired) in a small bowl. Set aside.

In a medium bowl, combine the garbanzo beans, grapes, feta cheese, onion, mint, oregano, salt, and pepper.

Place a leaf of lettuce in or on each piece of pita bread, followed by slices of cold lamb and a heaping spoonful of the prepared salad mix. Top with tahini dressing. Serve immediately.

# Seared Salmon with Spinach and Grapes

This is a "power food" lover's dream. Grilled fillets of salmon are served over a bed of wilted spinach and topped with a savory grape and red wine sauce. It is a nutritious combination that will leave you feeling both satisfied and slightly virtuous.

SERVES 4

| | |
|---|---|
| 1 | TEASPOON SALT |
| 1 | TEASPOON DRY MUSTARD |
| 1 | TEASPOON DRIED THYME (1 TABLESPOON FRESH) |
| ½ | TEASPOON PEPPER |
| FOUR | 6-OUNCE SALMON STEAKS OR FILLETS |
| 2 | TEASPOONS HONEY |
| 3 | TEASPOONS OLIVE OIL, DIVIDED |
| 1 | LARGE BUNCH SPINACH, WASHED AND STEMMED |
| 1 | CLOVE GARLIC, MINCED |
| 2 | CUPS RED GRAPES, HALVED |
| ½ | CUP DRY RED WINE |

Preheat oven to 325 degrees F.

Combine the salt, mustard, thyme, and pepper in a small bowl. Drizzle the salmon fillets with the honey and sprinkle with the seasoning. Reserve any remaining seasoning mixture.

Heat 2 teaspoons of the olive oil in a nonstick skillet or sauté pan. Brown both sides of the salmon fillets over medium-high heat, about 4 minutes per side. While the salmon is browning, toss the cleaned spinach and garlic with the remaining 1 teaspoon olive oil and arrange in a baking dish. Place the browned salmon on the bed of spinach, cover loosely with aluminum foil, and bake in the preheated oven for 8 to 10 minutes, or until the salmon is just cooked through.

Meanwhile, sauté the grapes for one minute over medium-high heat in the skillet used to brown the salmon. Add the wine, bring to a boil, and reduce the quantity by half. Season the sauce to taste with the remaining spice mixture. Serve the salmon on the wilted spinach, topped with the grape and wine sauce.

PB&G's

::::::: kids

:::::: Kids should learn to cook. It's a simple fact that the habits children learn stay with them throughout their lives.
If your ten-year-old can make her own snacks or help prep dinner, she'll be learning the skills and the habits to make her own meals for the rest of her life.
Teaching children the pleasures of cooking is a wonderful gift that they will use and enjoy for the rest of their lives.
Start with grapes, a food they already love; add some fun with these tasty recipes; and get them started on learning to love the kitchen.

Fresh Trail Mix

# Frosty Fresh Grape Pops

These are treats you won't object to when your kids are dying for something sweet. Make these in fancy molds or simply use paper cups. They can be made with or without grape pieces, depending on your preference.

| MAKES | 4½ | CUPS GREEN, RED, AND/OR BLACK GRAPES, PLUS 2 CUPS FRESH GRAPES, |
|---|---|---|
| 4 JUICE POPS | | HALVED (OPTIONAL) |
| OR 8 POPS | 2 | TEASPOONS SUGAR |
| WITH GRAPES | | FOOD COLORING (OPTIONAL) |

Juice the 4½ cups of grapes in a vegetable juicer or puree the grapes in a blender or food processor until smooth. With a plastic spatula, press the mixture through a fine strainer into a mixing bowl. Discard the solids. Skim any foam off the top. Dissolve the sugar in the juice. Add a few drops of red, green, or purple food coloring if desired.

To make the juice pops, fill molds or 5-ounce paper cups with the juice, skimming any foam off the tops. If using cups, cover each cup with foil and poke a wooden stick through the center of the foil for a handle. If using fruit pieces, put about ¼ cup halved grapes into each mold. Pour in the juice and skim any foam off the top. Place the molds or cups in the freezer for at least 2 hours, or until frozen. To remove the pops from the cups, take off the foil and tear away the paper.

Variation: To make layered juice pops, make 2 batches of grape ice pop mixture from different colored grapes and food coloring if desired. Fill the molds half full with one color and freeze. When hard, fill the remainder of the molds with the other color and freeze.

# Grape Spread with Cream Cheese and Nuts

This is a gooey grape spread that is great to keep on hand for both kids and adults. Enjoy it on bagels, crackers, or tortillas. Use it as a dip for celery sticks, carrots, and apple slices, or as the perfect tea sandwich filling for a spontaneous teddy-bear tea party. Neufchâtel or reduced-fat cream cheese are fine substitutes.

MAKES 1½ CUPS

1   BLOCK (8 OUNCES) CREAM CHEESE, SOFTENED
¼   CUP CHOPPED NUTS AND/OR SEEDS SUCH AS PECANS, ALMONDS, PEANUTS, PISTACHIOS, WALNUTS, MACADAMIAS, OR SUNFLOWER OR PUMPKIN SEEDS
½   CUP GREEN, RED, AND/OR BLACK GRAPES, CHOPPED

Stir together the cream cheese, nuts, and grapes in a small bowl.
Keep refrigerated for up to 3 days.

# Magic Wands

Skewering an assortment of sweet and savory tidbits helps make a balanced meal look more like a treat. These skewers are topped with slices of ripe star fruit to make them a magical meal. If you are concerned about younger kids, snip off the point of the skewers before adding the star fruit, to ensure that no one gets poked.

SERVES 6

36   GREEN, RED, AND/OR BLACK GRAPES
1   CUP BITE-SIZED PIECES OF HAM, SALAMI, OR PEPPERONI
3   PIECES OF STRING CHEESE, CUT INTO ½-INCH LENGTHS
12 OR 18   PIECES OF CHEESE TORTELLINI, COOKED AND COOLED
6   EXTRA LONG, HEAVY-DUTY WOODEN SKEWERS (AVAILABLE AT MANY GOURMET SHOPS)
1   STAR FRUIT CUT INTO 6 SLICES

Thread the grapes, meats, cheese, and tortellini decoratively onto the skewers. Top with a slice of star fruit. Keep chilled until ready to serve.

Grape Spread
with Cream
Cheese and Nuts

Hummus and
Grape Wraps

## Hummus and Grape Wraps

These vegetarian wraps are a light and healthy meal that even picky eaters will devour. Make extra so you can keep them in the fridge for grab-and-go meals or a school lunch.

SERVES 4

- 4 FLOUR TORTILLAS (IF YOU CAN GET YOUR KID TO EAT THE GREEN OR BROWN ONES, GO FOR IT!)
- ½ CUP HOMEMADE OR COMMERCIALLY PREPARED HUMMUS
- 4 SMALL ROMAINE LETTUCE LEAVES
- 4 SLICES PROVOLONE OR HAVARTI CHEESE
- 1 CUP GREEN, RED, AND/OR BLACK GRAPES, HALVED

Spread the hummus evenly on the tortillas. Top each with a lettuce leaf, a slice of cheese, and a sprinkling of grapes. Roll them up like a burrito and serve, or wrap in plastic wrap and keep chilled until ready to serve.


## Fresh Trail Mix
( PICTURED ON PAGE 101)

Kids love finger foods and snack mixes. With the creation of so many insulated lunch carriers and cold packs, there is little reason to rely on processed snack bars or dried trail mix for a pick-me-up, even when you are far away from your kitchen or fridge.

SERVES 2 TO 4

- 1 CUP GREEN, RED, AND/OR BLACK GRAPES
- 4 OUNCES JACK OR CHEDDAR CHEESE, DICED
- ½ CUP CHOCOLATE OR CAROB CHIPS
- ½ CUP ALMONDS, PEANUTS, OR SUNFLOWER SEEDS

Combine the ingredients in resealable bags. Keep chilled until ready to serve.

## PB&G'S
( PICTURED ON PAGE 98 )

Cut out a little more sugar from your family's diet by introducing them to the glories of Peanut Butter and Grapes. One taste and you may never go back to jars of sticky jelly. Spread peanut butter on toast, bagels, tortillas, or crackers and sprinkle generously with sliced fresh grapes.

## EDIBLE JEWELRY

This is a great way to promote healthy snacking while keeping the kids busy. Try it on the next rainy day or whenever an interactive project is needed.

Arrange fresh green, red, and black grapes and assorted cereals, such as Kix and Cheerios, in individual bowls. Tie a piece of cereal to one end of a necklace- or bracelet-length piece of waxed, unflavored dental floss and thread the other end with a large, blunt-tipped needle. Let the kids thread the fruit and cereal onto the dental floss to create their own unique and edible jewelry. Help them out by tying knots and encouraging them to eat their creations and start again. The dental floss easily cuts through the grapes when they eat them, so there is little dripping. Be sure to suggest they use the dental floss after they are finished snacking!

# Alien Eyeball Soup

Can't you hear the kids already? "Yikes! What is this jiggly, fizzy green glop? Alien eyeballs . . . AAAAAAhhhhh!" If you are an especially devoted entertainer, you will take the time to peel a few grapes, because it is remarkable how much a peeled black grape can look like an alien eyeball! To reduce the sugar, make this with sugar-free gelatin and soda and no chocolate. It will still inspire shouts and groans of glee.

| SERVES 8 | ONE | 6-OUNCE PACKAGE GREEN GELATIN |
|---|---|---|
| | 2 | CUPS GREEN, RED, AND/OR BLACK GRAPES |
| | ¼ | CUP WHITE AND/OR SEMI-SWEET CHOCOLATE CHIPS (OPTIONAL) |
| | 1 | LITER LEMON-LIME SODA |

Make the gelatin according to the manufacturer's directions.

Peel some of the grapes to make them more slippery and unusual. Poke dark chocolate chips into the stem ends of the green grapes and white chocolate chips into the stem ends of the dark grapes to look like pupils. Stir to combine the gelatin and grapes. Spoon into individual bowls or serve in a soup terrine. Pour on the soda and serve while fizzing.

Black Fruits in Lavender-Honey Glaze

::::::: finish

There aren't that many foods that are both great to eat
and really good for you. Chocolate cake isn't a health food,
and rice cakes don't appear on too many dessert menus.
Grapes, however, have a special place. They're sweet,
juicy, crisp, and flavorful. By themselves or in one of these
desserts, they're a sweet ending to your meal.
Today, scientists around the world are finding links
between fresh grapes and the potential to fight human
illnesses such as cancer and heart disease.
Preliminary research has connected consuming grapes with
reduced risk of diabetes and arthritis. Modern science is
verifying the fabled health benefits of the fruit of the vine.

Black Grape Sorbet

# Black Fruits in Lavender-Honey Glaze

( PICTURED ON PAGE 110 )

This dish is best served in the height of summer, when the fruit is perfect and the lavender is still blooming. Be sure to use only lavender blossoms that are food-grade and not treated with pesticides or preservatives.

SERVES 6

| | |
|---|---|
| ¼ | CUP HONEY, WARMED |
| ¼ | CUP FRESH ORANGE JUICE |
| 1½ | TEASPOONS GRATED ORANGE ZEST |
| 1 | TEASPOON DRIED LAVENDER BLOSSOMS |
| 2 | CUPS BLACK GRAPES |
| 1½ | CUPS BLACKBERRIES |
| 1½ | CUPS DARK, SWEET CHERRIES, PITTED |
| 1 | CUP BLACK-SKINNED PLUMS, SLICED |
| 6 | SPRIGS FRESH LAVENDER FOR GARNISH |

In a small bowl, whisk together the warm honey, orange juice, orange zest, and lavender blossoms. Set aside.

In a large bowl, combine the grapes, blackberries, cherries, and plum slices. Pour the honey mixture over the fruits and toss gently to coat. Spoon the fruit and honey into individual serving bowls and garnish with fresh lavender. Serve immediately.

# Fresh Fruit Tart

A perfect fruit tart can be as simple as fresh grapes encased in a frame of puff pastry. If using only red or black grapes, you can use red currant jelly to add a ruby glint. Green grapes or mixed colors look best with apple jelly.

SERVES 4 TO 6

- 1 SHEET FROZEN PUFF PASTRY, THAWED
- 1 EGG, BEATEN
- 1 TABLESPOON SUGAR
- 2 CUPS GREEN, RED, AND/OR BLACK GRAPES, OR A SELECTION OF GRAPES AND FRESH BERRIES
- 2/3 CUP APPLE JELLY

Preheat oven to 425 degrees F.

Roll the pastry out on a lightly floured surface to 12 by 9 inches. Cut so you have a 12-by-6-inch rectangle and a 12-by-3-inch strip.

Place the larger piece of pastry on a baking sheet. Brush the top edge with some of the egg. Cut the strip of pastry into three 1-by-12-inch lengths. Cut one strip in half and then arrange the strips on the larger piece of pastry to form a 1-inch "frame" on top of the base. Press down lightly. Brush the top frame with more egg and cut a decorative crisscross pattern onto the pastry with the back of a knife. Sprinkle with the sugar.

Prick the exposed base well with a fork to prevent it from puffing. Bake in the center of the preheated oven for 12 to 15 minutes, or until the pastry is a dark golden brown. If the center puffs or bubbles, simply pierce the pastry again with a fork or gently press it to flatten. Cool.

Arrange the grapes in the pastry case. Melt the jelly in a small saucepan or microwave. Using a pastry brush, generously dab the jelly over the grapes to coat. Serve plain or with a dollop of cream or ice cream.

# Grape and Blackberry Crisp

A crisp, sweet oat topping over fruits and berries makes the quintessential summer dessert. This can also be baked in individual ramekins for a more elegant presentation. Baking time will be shortened to 25 minutes. Serve warm with a scoop of ice cream.

| TOPPING | 1¼ | CUPS ALL-PURPOSE FLOUR |
|---|---|---|
| | ¾ | CUP BROWN SUGAR, LIGHTLY PACKED |
| | ½ | CUP ROLLED OATS |
| | ½ | CUP (1 STICK) BUTTER |
| | 1 | TEASPOON VANILLA EXTRACT |
| | ¼ | TEASPOON SALT |
| | | |
| SERVES 6 | 3 | CUPS BLACK GRAPES |
| | 3 | CUPS BLACKBERRIES, FRESH OR FROZEN |
| | ¼ | CUP SUGAR |
| | 1 | TABLESPOON CORNSTARCH |

Preheat oven to 350 degrees F. Spray a 9-inch square baking pan generously with baking spray.

Make the topping by mixing the flour, brown sugar, oats, butter, vanilla, and salt until crumbly and pea-sized pieces form. Set aside.

In a large bowl, combine the grapes, blackberries, sugar, and cornstarch. Pour into the prepared pan. Sprinkle the crumb topping evenly over the fruit. Bake in the middle of the preheated oven for 40 to 45 minutes, until the top is golden brown and the fruit is bubbling in the center. Serve warm with whipped cream or iced cream.

# Chocolate Cabernet Torte

Inspired by the classic Black Forest cake, this may just become your new favorite dessert. It's not too sweet, which makes it the perfect finish for a multicourse meal. It's pretty easy to make from scratch, but for a lightning-quick version, use your favorite boxed cake mix. Cabernet and chocolate make a great combination, but if you want to splurge, use a good tawny port. Commercial whipped cream stabilizers, which are very popular in Europe, are available in the baking section of many supermarkets and will make a firmer cream filling in case the cake needs to be transported.

SERVES 8

| | |
|---|---|
| 2 | CUPS HEAVY CREAM |
| 2½ | CUPS (1 POUND) RED AND/OR BLACK GRAPES, SLICED |
| 3 | LAYERS OF CHOCOLATE CAKE (SEE RECIPE ON FACING PAGE) |
| ½ | CUP CABERNET SAUVIGNON |
| | ADDITIONAL GRAPES FOR GARNISH |

Whip the cream until stiff peaks form. In a medium bowl, stir together two-thirds of the whipped cream and the sliced grapes.

Place a cake layer on a serving plate and carefully spoon on 2 to 3 tablespoons of the wine to evenly moisten the cake. Spread half of the grape mixture over the prepared cake layer and top with another round of cake. Repeat with more wine and the remaining grapes and cream mixture. Top with the final cake layer and the remaining wine and finish with the remaining one-third of the plain whipped cream.

Garnish with additional grapes and keep chilled until ready to serve.

| BASIC | 1⅔ | CUPS CAKE FLOUR |
|---|---|---|
| CHOCOLATE | ¼ | CUP DUTCH COCOA |
| CAKE | 1 | TEASPOON BAKING SODA |
| | ½ | TEASPOON SALT |
| MAKES 3 | ½ | CUP (1 STICK) UNSALTED BUTTER, |
| | | AT ROOM TEMPERATURE |
| 8-INCH | 1½ | CUPS SUGAR |
| CAKE LAYERS | 2 | EGGS |
| | 1½ | TEASPOONS VANILLA EXTRACT |
| | 2 | OUNCES UNSWEETENED CHOCOLATE, MELTED |
| | ¾ | CUP MILK |

Spray three 8-inch round cake pans generously with baking spray. Preheat oven to 350 degrees F.

Sift together the flour, cocoa, baking soda, and salt. With an electric mixer, cream together the butter and sugar until light-colored and fluffy. Add the eggs one at a time, beating thoroughly between each addition. Stir in the vanilla and unsweetened chocolate. Fold in half of the dry ingredients, then half of the milk, then the remaining dry ingredients and milk. Stir until evenly mixed.

Pour the batter evenly into the prepared pans. Bake for 12 minutes, or until the cake has pulled away slightly from the sides of the pans and a toothpick poked in the center comes out clean.

Let cool for 3 minutes and then turn out onto a cooling rack. Cool completely.

# Grape and Cranberry Tea Bread

A dainty moist loaf perfect for brunch or afternoon tea. If you are watching your calories, use low-fat sour cream and skim milk. This freezes well if you would like to have a loaf ready for surprise guests.

| MAKES | | |
|---|---|---|
| 2 LOAVES | 2½ | CUPS ALL-PURPOSE FLOUR |
| | ¾ | CUP SUGAR |
| | 1 | TABLESPOON BAKING POWDER |
| | 1 | TEASPOON SALT |
| | ¼ | CUP VEGETABLE OIL |
| | 1 | CUP SOUR CREAM |
| | ¼ | CUP MILK |
| | 1 | EGG, BEATEN |
| | 2 | TEASPOONS FINELY GRATED LEMON ZEST |
| | 1½ | CUPS RED OR BLACK GRAPES, HALVED |
| | ½ | CUP DRIED CRANBERRIES |
| | ½ | CUP WALNUTS, CHOPPED (OPTIONAL) |

Preheat oven to 350 degrees F. Spray two 9-by-5-inch loaf pans generously with baking spray.

In a large bowl, stir together the flour, sugar, baking powder, and salt. Make a well in the center of the dry ingredients and add the oil, sour cream, milk, egg, and lemon zest. Stir, gradually combining the wet and dry ingredients. Fold in the grapes, cranberries, and walnuts, if desired. Pour the batter evenly into the prepared pans. Bake for 40 to 45 minutes, or until a wooden pick inserted near the center comes out clean. Cool for 5 minutes, then turn the loaves out onto a wire rack. Cool completely before slicing.

# Coconut Tapioca with Grapes

Tapioca with coconut milk is a favorite snack throughout Asia. It isn't too sweet, it's easy to make, and it can be enjoyed warm on chilly days or icy cold when the sun is blazing. If you've never tried cold tapioca with grapes on a hot and humid day, you are in for a treat.

SERVES 6

½ CUP TAPIOCA PEARLS (NOT THE INSTANT KIND FOUND AT MOST SUPERMARKETS)

1½ CUPS WATER

⅓ CUP SUGAR

PINCH OF SALT

½ CUP COCONUT MILK (FOR BEST RESULTS, SPOON OFF THE THICK, RICH CREAM THAT SETTLES AT THE TOP OF A CAN OF COCONUT MILK)

1 CUP GREEN, RED, AND/OR BLACK GRAPES, HALVED

GINGER SYRUP (OPTIONAL; RECIPE FOLLOWS)

Soak the tapioca pearls in plenty of water for 15 to 20 minutes. Drain and rinse well. In a medium saucepan, combine the tapioca with the water, sugar, and the salt. Bring the mixture to a boil, stirring constantly, then reduce the heat to low and simmer until the tapioca is tender, 5 to 8 minutes. Remove from the heat and stir in the coconut milk. For a warm pudding, serve in bowls topped with the cold grapes. Alternatively, chill and serve cold topped with the grapes and a drizzle of fresh ginger syrup.

GINGER SYRUP

MAKES ½ CUP

½ CUP WATER

½ CUP SUGAR

2 TO 3 QUARTER-SIZED SLICES OF FRESH GINGER

Place the water, sugar, and ginger in a small saucepan and bring to a boil. Boil for 5 minutes and cool. Discard the ginger and chill. Will keep refrigerated for 3 weeks.

# Grape and Apricot Clafouti

A clafouti is a traditional French dessert. It is remarkably quick and simple to prepare, so ask your kids to help. This version is more custardlike than some, and like all truly great desserts, this one is wonderful for breakfast the next day. Use half-and-half in place of the milk for a creamier and richer custard.

| | | |
|---|---|---|
| SERVES | 1 | CUP GREEN, RED, AND/OR BLACK GRAPES |
| 6 TO 8 | 8 TO 10 | APRICOT HALVES, SLICED (CANNED IS FINE) |
| | 3 | EGGS |
| | ⅔ | CUP MILK |
| | ½ | CUP SUGAR |
| | ⅓ | CUP FLOUR |
| | ½ | TEASPOON VANILLA EXTRACT |
| | | PINCH OF SALT |
| | | POWDERED SUGAR FOR GARNISH |

Preheat oven to 325 degrees F. Butter a 9-inch fluted tart dish or pie plate.

Arrange the grapes and apricot slices in the bottom of the dish. In a medium bowl, whisk together the eggs, milk, sugar, flour, vanilla, and salt. Pour over the prepared fruit and bake in the middle of the oven for 50 minutes, or until the clafouti is firm to the touch and lightly browned. Cool slightly. Dust the top generously with powdered sugar and serve warm.

# Health Nut Muffins

These little gems are jam-packed with just about everything that is good for you. They are low in fat and full of fiber. If you end up with leftovers, use them in the Breakfast Trifles (page 91).

| | | |
|---|---|---|
| MAKES 12 | 1 | CUP ALL-PURPOSE FLOUR |
| SMALL MUFFINS | ¾ | CUP QUICK-COOKING OATS |
| | ½ | CUP BROWN SUGAR, LIGHTLY PACKED |
| | ¼ | CUP WHEAT BRAN |
| | 1 | TABLESPOON BAKING POWDER |
| | ½ | TEASPOON GROUND CINNAMON |
| | ½ | TEASPOON GROUND ALLSPICE |
| | ¼ | TEASPOON SALT |
| | ¾ | CUP GREEN, RED, AND/OR BLACK GRAPES, HALVED |
| | ½ | CUP GRATED CARROTS |
| | ½ | CUP SKIM MILK |
| | ⅓ | CUP APPLESAUCE |
| | 3 | TABLESPOONS VEGETABLE OIL |
| | 1 | EGG, BEATEN |
| | ½ | CUP (2 OUNCES) WALNUTS, CHOPPED, FOR GARNISH |

Preheat oven to 350 degrees F. Line 12 muffin cups with paper liners.

Combine the flour, oats, brown sugar, bran, baking powder, cinnamon, allspice, and salt in a medium bowl and mix well. Stir in the grapes, carrots, milk, applesauce, oil, and egg until evenly moistened.

Spoon the batter evenly into the prepared muffin tins and sprinkle with the walnuts.

Bake for 20 to 25 minutes, or until a wooden toothpick comes out clean. Cool. Store in an airtight container.

# Grapes Suzette

The title alone is almost too good to pass up. But a better reason to make these is that crêpes drenched in sweet orange butter, grapes, and cognac are simply ephemeral.

SERVES 4

¼ CUP (½ STICK) BUTTER
¼ CUP SUPERFINE SUGAR
½ CUP FRESH ORANGE JUICE
1 TEASPOON GRATED ORANGE ZEST
1 CUP GREEN, RED, AND/OR BLACK GRAPES, SLICED
8 SWEET CRÊPES (RECIPE FOLLOWS)
¼ CUP COINTREAU, GRAND MARNIER, OR COGNAC (OPTIONAL)

Melt the butter in a large skillet or sauté pan. Stir in the sugar, orange juice, orange zest, and sliced grapes. Cook on medium-high until bubbly. Reduce the heat to low. Working quickly, place a crêpe into the sauce and fold into a fan shape. Lift the crêpe from the pan and arrange on a warm serving dish or platter. Repeat with the remaining crêpes. Increase the heat to medium-high and add the Cointreau, being careful to avoid lighting the liquor on fire unless that is a desired effect. Stir to blend and spoon the warm grape sauce over the prepared crêpes. Serve immediately.

SWEET CRÊPES

MAKES 8 TO
10 CRÊPES

1 CUP ALL-PURPOSE FLOUR
¼ CUP SUGAR
PINCH OF SALT
2 WHOLE EGGS PLUS 1 EGG YOLK
1 CUP MILK
1 TABLESPOON BRANDY OR COGNAC
1 TABLESPOON VEGETABLE OIL

In a medium bowl, combine the flour, sugar, and salt. Make a well in the middle of the dry ingredients and add the eggs, yolk, milk, and brandy. Gradually blend the dry ingredients into the wet ingredients until the mixture is smooth and lump-free. Let the batter rest for at least 30 minutes.   >>>

When you are ready to cook the crêpes, the batter will have thickened slightly from resting. Add water to the batter 1 tablespoon at a time until it has thinned to the consistency of heavy whipping cream.

Heat a nonstick skillet or sauté pan over medium-high heat. Brush lightly with the oil and swirl in a scant ¼ cup of the crêpe batter to coat the bottom of the pan. Cook for 1 minute, or until the very edges of the crêpe are lightly brown. Flip with a flexible spatula and brown the other side. Repeat with the remaining batter. Use crêpes immediately or cover and keep chilled for up to 24 hours.

## Black Grape Sorbet
( PICTURED ON PAGE 113 )

This is a refreshing palate cleanser to enjoy between courses of a fancy meal or as a healthy, bright dessert that is naturally low in calories. Actually, any color of grapes works for this, but the black grapes make an especially pretty sorbet. Adding a tablespoon of vodka softens and improves the texture.

SERVES 4

4  CUPS BLACK GRAPES

¼  CUP SUGAR

½  TEASPOON LEMON ZEST

1  TABLESPOON FRESH LEMON JUICE

1  TABLESPOON VODKA (OPTIONAL)

Puree the grapes, sugar, lemon zest, and lemon juice in a blender until smooth. Pour into a saucepan and bring to a boil over high heat. Boil for 30 seconds to infuse the flavor of the grape skins and lemon zest. Remove from the heat and strain through a fine-mesh sieve. Discard the solids. Stir the vodka, if desired, into the puree and cool completely. Freeze in an ice-cream maker, according to the manufacturer's instructions, or pour into a shallow, nonreactive pan and put in the freezer. Stir every 10 to 15 minutes until the sorbet is frozen.

# Index

# Table of Equivalents

The exact equivalents in the following tables have been rounded for convenience.

**LIQUID / DRY MEASURE**

| U.S. | METRIC |
|---|---|
| ¼ teaspoon | 1.25 milliliters |
| ½ teaspoon | 2.5 milliliters |
| 1 teaspoon | 5 milliliters |
| 1 tablespoon (3 teaspoons) | 15 milliliters |
| 1 fluid ounce (2 tablespoons) | 30 milliliters |
| ¼ cup | 60 milliliters |
| ⅓ cup | 80 milliliters |
| ½ cup | 120 milliliters |
| 1 cup | 240 milliliters |
| 1 pint (2 cups) | 480 milliliters |
| 1 quart (4 cups, 32 ounces) | 960 milliliters |
| 1 gallon (4 quarts) | 3.84 liters |
| 1 ounce (by weight) | 28 grams |
| 1 pound | 448 grams |
| 2.2 pounds | 1 kilogram |

**LENGTHS**

| U.S. | METRIC |
|---|---|
| ⅛ inch | 3 millimeters |
| ¼ inch | 6 millimeters |
| ½ inch | 12 millimeters |
| 1 inch | 2.5 centimeters |

**OVEN TEMP**

| FARENHEIT | CELSIUS | GAS |
|---|---|---|
| 250 | 120 | ½ |
| 275 | 140 | 1 |
| 300 | 150 | 2 |
| 325 | 160 | 3 |
| 350 | 180 | 4 |
| 375 | 190 | 5 |
| 400 | 200 | 6 |
| 425 | 220 | 7 |
| 450 | 230 | 8 |
| 475 | 240 | 9 |
| 500 | 260 | 10 |